SCIENCE
A CLOSER LOOK

BUILDING SKILLS
Math

Mc Graw Hill **Macmillan McGraw-Hill**

Instructions for Copying

Answers are printed in non-reproducible blue. Copy pages on a light setting in order to make multiple copies for classroom use.

Contents

LIFE SCIENCE

Chapter 1 Plants Are Living Things 1
Chapter 2 Plants Grow and Change 3
Chapter 3 All About Animals 5
Chapter 4 Places to Live 7

EARTH SCIENCE

Chapter 5 Looking at Earth 9
Chapter 6 Caring for Earth 11
Chapter 7 Weather and Seasons 13
Chapter 8 The Sky 15

PHYSICAL SCIENCE

Chapter 9 Matter Everywhere 17
Chapter 10 Changes in Matter 19
Chapter 11 On the Move 21
Chapter 12 Energy Everywhere 23

Graphing in Science

Tally Charts 27
Picture Graphs 29
Bar Graphs 31

Math Strategies

Number Sense 33
Algebra and Functions 35
Measurement and Geometry 37
Statistics, Data Analysis, and Probability 39
Mathematical Reasoning 41

Introduction and Rationale

Math was designed to help introduce students to doing math activities within a science framework. Students will learn how to use mathematics in real-world situations. Scientists use math every day when conducting experiments—to count the number of animals in an area, measure the amounts to use, or figure out the probability of an event. Students will begin to discover how to analyze and describe nature mathematically.

Every stage of the scientific process helps students develop thinking and problem-solving strategies. Students will learn to:

- Collect and organize data.

- Ask questions and form hypotheses.

- Conduct experiments.

- Use observation and logical reasoning to determine whether or not the experiments' results support their hypotheses.

- Create and record the results of their experiments in graphs, charts and tables.

The *Math* worksheets guide the use of *Math in Science* extensions in each chapter of the Student Edition. The *Graphing in Science* pages introduce grade-appropriate graphing skills in a science context. The *Math Strategies* pages offer practice in the five Content Standards from the National Council of Teachers of Mathematics, including the following three examples:

- Number Operations, comparing and ordering integers and solving problems that require arithmetic operations with whole numbers and fractions;

- Algebra, writing and solving problems involving rate, average speed, distance, and time;

- Reasoning and Proof, using counters and drawings to estimate unknown quantities.

Seeds of All Sorts!

Michael sorted his seeds. He made a picture graph to show how many of each seed he has.

beans	🫘 🫘 🫘 🫘 🫘
sunflower seeds	🌻🌻🌻🌻 🌻🌻🌻
corn kernels	🌽🌽🌽🌽🌽
peas	🟢🟢🟢🟢🟢 🟢🟢🟢🟢

Read a Graph

Does Michael have more sunflower seeds or beans?

(sunflower seeds)   beans

Remember

A picture graph helps you solve a problem.

Name _____

Write a number sentence to show how you know.

_ _

7 − 5 = 2

If he found 6 more peas, how many would he have? _____ 15

Write a number sentence to show the answer.

_ _

9 + 6 = 15

Try it Again

Ana sorted seeds. She made a picture graph to show how many of each seed she has.

Ana's Seeds								
apple seeds								
melon seeds								
orange seeds								

How many orange seeds does Ana have? _____ 7

How many seeds are not orange seeds? _____ 11

Write a number sentence to show how you know.

_ _

8 + 3 = 11

Fruit for Sale

Jasmine wanted to buy some fruit at the market. She had these coins.

Remember

A quarter is worth 25 cents.
A dime is worth 5 cents.

Count the Coins

Can Jasmine buy an apple?

_no_

Can Jasmine buy a lemon and a lime?

_no_

How do you know?

Jasmine only has 70 cents.

Try It Again

Sanjay wanted to buy some vegetables at the market. He had these coins.

Can Sanjay buy a cucumber?

_____

yes

Can Sanjay buy a carrot and a tomato?

_____

no

How do you know?

_____

Sanjay only has 20¢.

Animal Graph

Tom made a bar graph to show what kinds of pets his friends have at home.

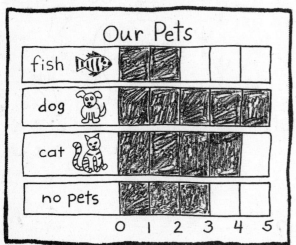

Make a Graph

Find out about your class's favorite pets. Make a bar graph to show which four pets your class likes best.

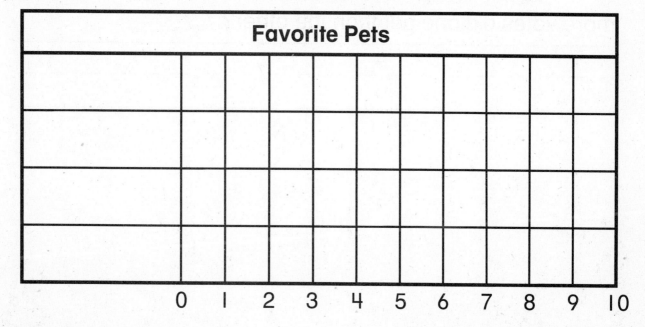

Try It Again

Look back at your Favorite Pets graph.
What did you find out?

Which pet got the most votes?
_____

- - - - - - - - - - - - - - -
Answers will vary.
_____

How many votes did that pet get?

_____

Which pet got the fewest votes?
_____

- - - - - - - - - - - - - - -
Answers will vary.
_____

Pick two pets from your graph. How many
more votes did one get than the other?
_____

- - - - - - - - - - - - - - -
Answers will vary.
_____

Count the Legs

Some animals use their legs to find food. Animals can have different numbers of legs.

Remember

Start with the smallest number of legs.

Put Them In Order

Count the number of legs each animal has. Put the animals in order from the smallest number of legs to the largest number of legs.

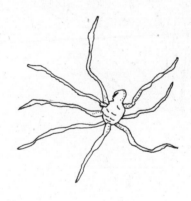

Name _____

Try It Again

These animals all live in a lake habitat. Count the number of legs each animal has. Put the animals in order from the smallest number of legs to the largest number of legs. Write 1, 2, and 3 on the lines. Write 1 next to the animal with the fewest legs, and write 3 next to the animal with the most legs.

2 _____

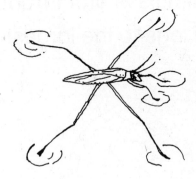

3 _____

1 _____

Adding Rocks

Peter sorted his rocks into two groups. Then he wrote a number sentence to show how many rocks he had all together.

Remember

A number sentence helps you solve a problem.

$$4 + 6 = 10$$

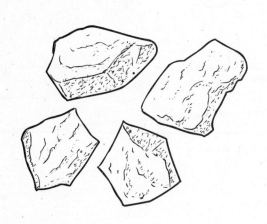
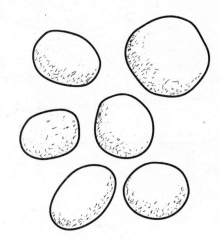

Write a Number Sentence

Collect your own rocks. Sort them into groups. Draw a picture of your groups. Write a number sentence to show how many rocks you have.

_____

Answers will vary, but should match drawings.

Name _____

Try It Again

Amy and Tom each collected rocks. They sorted their rocks into different groups. Write a number sentence for each picture.

_____

- -

2 + 7 = 9

_____

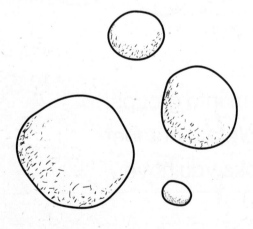 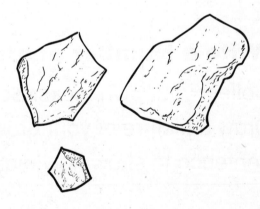

_____

- -

4 + 3 = 7

_____

Recycling Cans

John and his class picked up cans to recycle every day for one week. They made a picture graph to show how many cans they collected.

Cans We Collected

Days of week	Number of cans
Monday	🔋🔋🔋🔋🔋
Tuesday	🔋🔋🔋🔋
Wednesday	🔋🔋
Thursday	🔋🔋🔋🔋🔋🔋
Friday	🔋🔋🔋

Read a Graph

On which day did John's class collect the most cans?

_____

Thursday

_____

If the class got 5 cents for every can they collected, how much money would they have made on Monday?

_____

25 cents

_____

Name _____

Try It Again

Mr. Lee's class picked up cans to recycle every day for a week. They made a graph to show how many cans they collected.

Cans We Collected	
Days of week	Number of cans
Monday	🥫🥫🥫🥫
Tuesday	🥫🥫
Wednesday	🥫🥫🥫🥫🥫
Thursday	🥫🥫🥫
Friday	🥫🥫🥫

On which day did Mr. Lee's class find the fewest cans?

_____

Tuesday

On which day did Mr. Lee's class find the most cans?

_____

Wednesday

How many cans did the class find on Monday and Tuesday combined? Write a number sentence to show your answer.

_____

4 cans + 2 cans = 6 cans

Weather Graph

Anna asked her friends which activity they liked to do best on sunny days. She made a bar graph to show what she found out.

> **Remember**
>
> A bar graph needs a title.

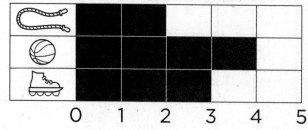

Sunny Day Activities

Make a Graph

Ask your classmates if they like to read a book, play a game, or draw a picture on a rainy day. Make a bar graph to show what you find out.

Rainy Day Activities										
read a book										
play a game										
draw a picture										

0   1   2   3   4   5   6   7   8   9   10

Name _____

Try It Again

Ask your classmates which type of weather they enjoy most. Fill in the bar graph to show your data.

	Favorite Weather			
6				
5				
4				
3				
2				
1				
	sunny	rainy	cloudy	snowy

Which weather got the most votes?

Answers will vary.

Which weather got the fewest votes?

Answers will vary.

How many children do not enjoy sunny weather?

Answers will vary.

Measure Time

Ricky estimated how long it would take him to do different activities. Then he used a clock to find out how long they really took. He made a chart.

Remember

You estimate when you guess how much time it takes to do something.

How long does it take?

Activity	Estimated time	Actual time
Brushing teeth	1 minute	2 minutes
Reading a book	20 minutes	30 minutes
Sleeping	16 minutes	540 minutes

Estimate Activities

Make a chart like Ricky's. Do your activities take longer than you estimated? How can you find out?

How long does it take?

Activity	Estimated time	Actual time

Name _____

Try It Again

Look back at your 'How long does it take?' chart. What did you find out?

Which estimate was the closest to the actual time?

_____

Answers will vary.

Which estimate was the farthest from the actual time?

_____

Answers will vary.

Which activity did you spend the most time doing?

_____

Answers will vary.

Which activity did you spend the least time doing?

_____

Answers will vary.

Weigh It

A scale measures weight.
Look at the scales below.
Put the fruits in order from
the lightest fruit to the
heaviest fruit.

Remember

Tools improve the accuracy
of estimates.

2 _____   1 _____   3 _____

Weigh Yourself

Estimate your weight. Then use a scale to
measure it. Check to see if you are right.

_____

My Prediction _Answers will vary._

_____

My Actual Weight _Answers will vary._

Name _____

Try It Again

Pick three classroom objects. Use a scale to weigh each object. Draw the objects in order from lightest to heaviest and record their weights on the lines. Draw the lightest object in the top box.

_____

_____

_____

Trail Mix Recipe

Carrie made trail mix. She used this recipe. She mixed everything together.

Remember

A number sentence helps you solve a problem.

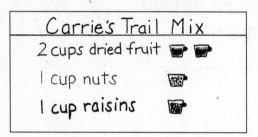

Carrie's Trail Mix
2 cups dried fruit
1 cup nuts
1 cup raisins

Write a Number Sentence

Make your own trail mix. Write a number sentence to show how many cups of each food you used in your mix.

_____

Number sentences will vary.

Try It Again

Ms. Lewis's class
made fruit salad
from this recipe.

Fruit Salad

2 cups sliced banana
1 cup grapes
2 cups sliced apples
1 cup berries

Write a number sentence to find how many
cups of fruit there are in all.

2 cups + 1 cup + 2 cups + 1 cup = 6 cups

Think of another fruit you would like to add
to the salad. How much would you add?

Answers will vary.

Write a number sentence to show how many
cups of fruit are in your new salad.

Answers will vary.

Comparing Magnets

Molly had two magnets. She wondered which one would pick up more paper clips. She compared the amounts.

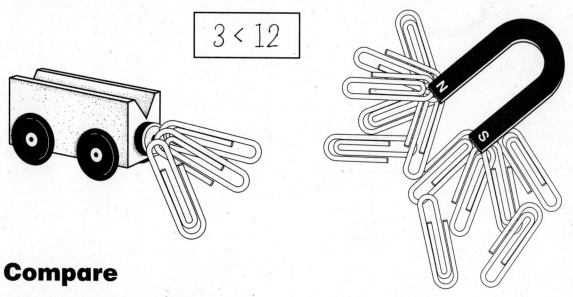

$3 < 12$

Compare

Use two different magnets. See which one picks up more paper clips. Compare the amounts.

_____

Number sentences will vary, but should accurately

_____

reflect a greater than or less than relationship.

Try It Again

Molly tried picking up paper clips with some other magnets. Compare the numbers.

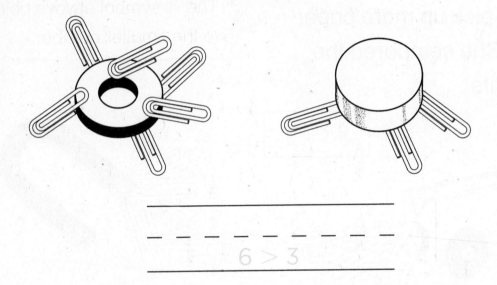

_____

6 > 3

_____

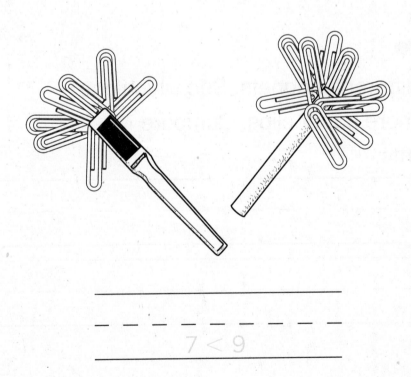

_____

7 < 9

_____

Stained Glass

Stained glass windows are made with many pieces of colored glass. When sunlight shines through stained glass, you can see different colors of light.

Remember

Use tally marks to keep track of the shapes you've already counted.

Sort the Shapes

What shapes do you see in the stained glass window above?

_____

circles, rectangles, ovals, triangles

How many circles do you see? _____ Answers will vary.

How many rectangles do you see? _____ Answers will vary.

Try It Again

Make your own stained glass design.

Use different shapes.

How many circles did you draw? _Answers will vary._

How many squares did you draw? _Answers will vary._

What other shapes did you draw? _____

Answers will vary.

Battery Chart

Sally did an experiment to find out which battery lasted the longest. She recorded her results in the chart below.

Remember

Subtract the smaller number from the larger one to find the difference.

Read a Chart

Which battery lasted the longest?

Battery B

My Batteries	
Battery	Hours
A	10 hours
B	15 hours
C	7 hours

How many more hours did battery B last than battery A? How do you know?

5 hours. I know this because 15 hours minus 10 hours equals 5 hours.

Name _____

Try It Again

Harold conducted his own battery experiment to see which battery lasted longest. Below are the results of his test.

Harold's Battery Experiment	
Battery	Hours
D	3 hours
E	7 hours
F	11 hours

Compare Sally's chart and Harold's chart.

Which batteries lasted the same amount of time?

_____

Battery C and Battery E

Which battery lasted the shortest time?

_____

Battery D

How could you find out how long all of the

batteries lasted together?

_____

I could add all the numbers in the charts.

Use with **Lesson 4**
Electricity

Tally Charts

A tally chart uses lines to show how many.

I stands for 1. 卅 stands for 5.

Mike's class made a tally chart of favorite ocean animals.

Use the tally chart to answer the questions.

Favorite Ocean Animals		
Animal	**Tally**	**Total**
(dolphin)	卅	5
(seal)	IIII	4
(shark)	卅 II	7
(crab)	卅	5

1. How many children chose the dolphin? _____ 5

2. Which animal got the fewest votes?

_____

seal

3. Which animals got the same amount of votes?

_____

crab   dolphin

Name _____

Practice It

Write each total.

Favorite Desert Animals

Animal	Tally	Total
	||||	4
	⊠||| |	6
	|||	3
	⊠||| ||	7

Ask 10 friends to pick a favorite forest animal.
Make a tally chart to show what you find.

Favorite Forest Animals

Animal	Tally	Total

Picture Graphs

A picture graph uses pictures to display amounts. Each picture stands for 1.

Kenya's class made a picture graph of favorite seasons. Most children chose summer.

Use the picture graph to answer the questions.

Favorite Seasons					
winter	❄	❄	❄		
spring	🌼	🌼	🌼	🌼	
summer	☀	☀	☀	☀	☀
fall	🍁	🍁	🍁	🍁	

1. How many children chose summer? ___5___

2. Which season got the fewest votes?

   _____

   winter

3. Which seasons got the same amount of votes?

   _____   _____

   spring   fall

Name _____

Practice It

Use the graph to answer the questions.

Each ☺ stands for 1.

Favorite Weather						
sunny ☀	☺	☺	☺	☺		
windy	☺	☺				
snowy ❄	☺	☺	☺	☺	☺	
stormy ☁	☺					

1. How many children chose windy weather? ___2___

2. Which weather got the fewest votes?

_____

_ _ _ _ _ _ _ _ _
stormy

_____

_ _ _ _ _ _ _ _ _

3. Which weather got the most votes? _____
snowy

Bar Graphs

A bar graph uses bars to display amounts.

Ali and her friends collected plastic bottles to recycle. She made a bar graph to show how many bottles each friend found.

Plastic Bottles Collected for Recycling

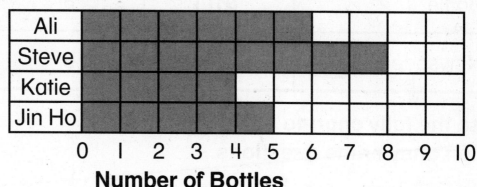

Children

Number of Bottles

1. How many bottles did Jin Ho collect? ____5____

2. Who collected the most bottles?

   _____

   _ _ _ _ _ _ _ _ _ _
   Steve

3. Who collected 1 more bottle than Katie?

   _____

   _ _ _ _ _ _ _ _ _ _
   Jin Ho

4. How many more bottles did Ali collect than Katie? ____2____

Name _____

Practice It

Mrs. Lena's class made a tally chart
to show how many items they recycled.

Item	Tally	Total
Can	ЦН ЦН	10
Bottle	ЦН I	6
Newspaper	ЦН III	8

**Use the tally chart to make a bar graph.
Then answer the questions.**

Items Collected for Recycling

Item											
	can										
	bottle										
	newspaper										

0 1 2 3 4 5 6 7 8 9 10

Number of Items

1. How many more newspapers than

   bottles were collected? _____ 2

2. How many cans and bottles were

   collected in all? _____ 16

Number and Operations
Compare Numbers

In science you may use large numbers.

Use these symbols to compare:

< means less than

> means greater than

= means equal to

Rob compared rainfall in his town for three years.

Rainfall for Three Years		
2001	2002	2003
21 inches	15 inches	25 inches

He wrote these numbers.

21 > 15   15 < 25   25 > 21

Write the missing symbols below.

21 ◯ 25   21 ◯ 15

Which year had the most rain? _____
2003

Practice It

Fill in the symbols. Solve.

1. Luis took the temperature outside. It was 64°F in the morning. It was 82°F in the afternoon.

   64° ◯ 82°

   _____

   When was it warmer? _the afternoon_____

2. On Monday it rained for one hour. On Tuesday it rained for 30 minutes.

   one hour ◯ 30 minutes

   Did it rain longer on Monday or Tuesday?

   _____

   _Monday_____

3. April had 17 sunny days. May had 22 sunny days. June had 18 sunny days.

   Which month had the most sunny days?

   _____

   _May_____

Algebra

Number Sentences

Number sentences help you add and subtract.
You can write number sentences to solve science problems.

Look at this problem.

> Each year has a fall, winter, spring, and summer.
> How many seasons are there in 2 years?

Think about the numbers. There are 4 seasons in one year. Add 4 plus 4 to find the total for 2 years:

$$4 + 4 = \underline{}8$$

How many seasons are there in 3 years?

Write a number sentence.

_____

– – – – – – – – – – – – – – – – – – –

4 + 4 + 4 = 12, 12 seasons

Practice It

Write a number sentence to solve.

1. A plant grew 6 inches in spring. It grew 3 inches in summer. Then it grew 2 inches in fall. How much did the plant grow in all?

   _____

   6 + 3 + 2 = 11, 11 inches

2. December 12 had 10 hours of sunlight. June 12 had 14 hours of sunilght. How much more sun was there on June 12?

   _____

   14 − 10 = 4, 4 hours

3. One summer had 19 rainy days. Fall had 12 rainy days. How many more days did it rain in summer?

   _____

   19 − 12 = 7, 7 days

Measurement

Compare Objects

You can measure objects to compare them. You can use tools or other objects to measure.

Look at these pencils.

Pencil A   **Pencil B**

How long is Pencil A? _____3_____ clips

How long is Pencil B? _____4_____ clips

Which pencil is longer? _____Pencil B_____

Practice It

Get some clips that are the same size.

Measure and compare each pair of objects.

1.

   car   truck

   How long is the car? _____ clips

   How long is the truck? __2__ clips

   _____

   – – – – – – – – – –

   Which toy is longer? the truck _____

2.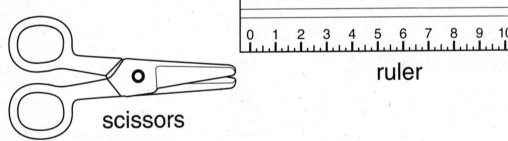

   ruler

   scissors

   How long are the scissors? __2__ clips

   How long is the ruler? __3__ clips

   _____

   – – – – – – – – – –

   Which tool is shorter? the scissors _____

Data Analysis and Probability
Sorting Leaves

You can sort plant parts. You can sort seeds by size. You can sort flowers by color. These leaves are sorted by shape.

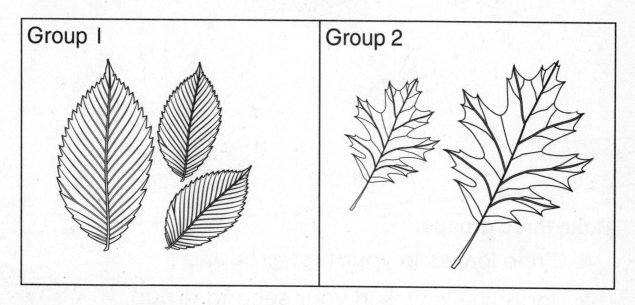

Group 1

Group 2

How are the leaves in Group 1 alike?

_____

_ _ _ _ _ _ _ _ _ _ _ _ _ _ _

Possible answer: The leaves are the same shape.

How are the leaves in Group 2 alike?

_____

_ _ _ _ _ _ _ _ _ _ _ _ _ _ _

Possible answer: The leaves are the same shape.

Practice It

How can you sort the leaves by shape?

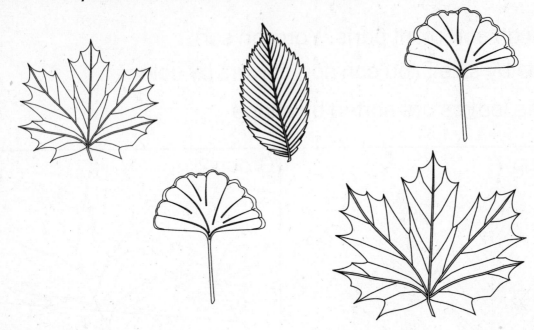

Make three groups.

▶ Circle leaves in your first group.

▶ Underline leaves in your second group.

▶ Shade leaves in your third group.

How are the leaves in each group alike?

_____

Responses will vary, but should indicate that child

_____

_____

observes shape differences between three types.

_____

Data Analysis and Probability

Reasoning and Proof

Use Pictures and Manipulatives

Sometimes in science, you must add or subtract. Use pictures and tools to help.

Look at this problem.

> Kelly hung a bird feeder. She saw 4 birds at the feeder in the morning. She saw 5 birds in the afternoon.

This picture shows what she saw.

morning   afternoon

These counters show what she saw.

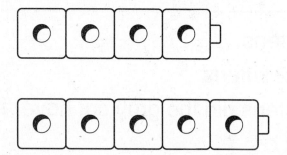

How many birds did Kelly see? ____9____

Name _____

Practice It

Solve it.

1. Mrs. Smith's class has a pet slug.

   The slug eats 2 carrots a week.

   How many carrots does it eat in 3 weeks? ___6___

Draw a picture to help.

Students' drawings should show six carrots, possibly in groups of two.

2. A gray cat had 5 kittens.

   An orange cat had 2 kittens.

   How many more kittens did the gray cat have

   than the orange cat? ___3___

Use counters to help. Students should use five counters, removing two for the answer.

Reasoning and Proof

Half-Inch Graph Paper

Graphing in Science
Math

Name _____

Two Centimeter Graph Paper